Natural Laboratories:
Scientists in
National Parks

DRY
TORTUGAS

Ruth A. Musgrave

Rourke
Educational Media

rourkeeducationalmedia.com

Before, During, and After Reading Activities

Before Reading: Building Background Knowledge and Academic Vocabulary

"Before Reading" strategies activate prior knowledge and set a purpose for reading. Before reading a book, it is important to tap into what your child or students already know about the topic. This will help them develop their vocabulary and increase their reading comprehension.

Questions and activities to build background knowledge:
1. Look at the cover of the book. What will this book be about?
2. What do you already know about the topic?
3. Let's study the Table of Contents. What will you learn about in the book's chapters?
4. What would you like to learn about this topic? Do you think you might learn about it from this book? Why or why not?

Building Academic Vocabulary

Building academic vocabulary is critical to understanding subject content.
Assist your child or students to gain meaning of the following vocabulary words.
Content Area Vocabulary
Read the list. What do these words mean?

- abundant
- commercial
- destruction
- endangered
- exoskeletons
- genetic
- indicators
- microscopic
- monitor
- pristine
- reserve
- techniques

During Reading: Writing Component

"During Reading" strategies help to make connections, monitor understanding, generate questions, and stay focused.
1. While reading, write in your reading journal any questions you have or anything you do not understand.
2. After completing each chapter, write a summary of the chapter in your reading journal.
3. While reading, make connections with the text and write them in your reading journal.
   a) Text to Self – What does this remind me of in my life? What were my feelings when I read this?
   b) Text to Text – What does this remind me of in another book I've read? How is this different from other books I've read?
   c) Text to World – What does this remind me of in the real world? Have I heard about this before? (News, current events, school, etc....)

After Reading: Comprehension and Extension Activity

"After Reading" strategies provide an opportunity to summarize, question, reflect, discuss, and respond to text. After reading the book, work on the following questions with your child or students to check their level of reading comprehension and content mastery.
1. What kinds of habitats and animals do scientists study at Dry Tortugas National Park? (Summarize)
2. Why are national parks important? (Infer)
3. What do scientists learn by tagging sharks or sea turtles? (Asking Questions)
4. What would you study if you were a scientist at Dry Tortugas National Park? (Text to Self Connection)

Extension Activity
Scientists often learn about animals through observation. Find a place to observe birds or other wild animals without them noticing. What behaviors do you see? How do they interact?

TABLE OF CONTENTS

CHAPTER ONE

DRY TORTUGAS NATIONAL PARK

Far out at sea, one of the most remote national parks in the United States features stunning views, crystal blue water, seven small islands, and **pristine** coral reefs and seagrass beds. The only way to get to Dry Tortugas National Park is by boat or seaplane.

Dry Tortugas National Park is in the Gulf of Mexico, 70 miles (113 kilometers) southwest of Florida. Most of the park is under water. The islands, also called keys, within the Dry Tortugas are the southernmost islands in the Florida Keys.

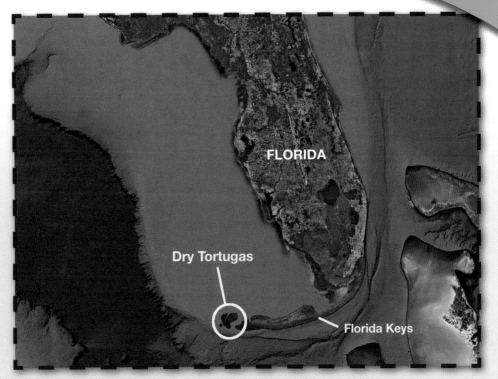

Dry Tortugas National Park is larger than Washington, D.C. The ocean is up to 82 feet (25 meters) deep within the park boundaries.

What Is a Key?

A key is a low island formed on top of an ancient coral reef. It is sometimes spelled cay or caye.

Fort Jefferson was built on Garden Key between 1846 and 1875. Today it serves as a historical site and visitor center.

Spanish explorer Juan Ponce de León discovered the islands in 1513. At the time, sea turtles were **abundant** on the beaches and the surrounding sea.

He named it *Las Tortugas*, which is Spanish for "the turtles." Later, mariners renamed it "Dry Tortugas" on nautical maps. That told sailors in search of drinking water that there was no fresh water on the islands.

Juan Ponce de León
1460 – July 1521

CHAPTER TWO

SCIENTISTS, THE ANIMALS, AND THE HABITAT

Sea turtles, corals, fish, sharks, and birds live in the Dry Tortugas. National parks like the Dry Tortugas provide unique living laboratories for scientists.

The Dry Tortugas is a crucial nesting site for sea turtles. Five species live or visit the area including loggerhead, green, leatherback, Kemp's ridley, and hawksbill.

loggerhead

green

"Since the 1800s, scientists have been studying and publishing information on natural resources, such as corals, reef fishes, and other marine life that abound in the coral reef ecosystems of the Tortugas region." — Natural Resource Report NPS/DRTO/NRR—2012/558, National Park Service

leatherback

Kemp's ridley

hawksbill

Unlike Ponce de Leon's time when sea turtles were abundant, all sea turtles species are now threatened or **endangered**. This is due to people overhunting sea turtles for their shells, meat, and eggs. Humans have also taken over many of the beaches sea turtles need to lay their eggs.

Coastal development, lights, and changes to the shoreline affect sea turtles' ability to survive.

Dr. Kristen Hart, research ecologist with the U.S. Geological Survey (USGS), studies the biology of sea turtles. She uses what she discovers to design programs and management strategies to help protect and save turtles.

Dr. Kristen Hart

Understanding nesting success helps biologists determine or predict the health of sea turtle populations. Biologists have been monitoring sea turtle nesting within the Dry Tortugas since 1980.

turtle tracks

turtle nest

Sea turtles leave a trail in the sand when they crawl on to shore to lay eggs.

Map labels:
- Gulf of Alaska
- North Pacific Ocean
- Canada
- Gulf of St. Lawrence
- United States
- North Atlantic Ocean
- Gulf of California
- Gulf of Mexico
- Bahamas
- Cuba
- Mexico
- Caribbean Sea
- Nicaragua

Traveling Turtles

Sea turtles that visit the Dry Tortugas migrate from Nicaragua, Northern Gulf of Mexico, Bahamas, and other places in the world.

A green sea turtle uses its back flippers to dig a nest in the sand.

After the turtle lays its eggs, it covers the nest with sand.

After the nest is buried, the turtle returns to sea. It does not guard the nest or raise the young.

During nesting season, female sea turtles come ashore to lay eggs. The female digs a nest in the sand and lays about 100 eggs the size of ping-pong balls. She covers the nest with sand and returns to sea.

Kristen and other biologists mark each nest, **monitor** the nesting sites, and wait for the baby turtles to hatch.

turtle tracks to
nesting site

turtle tracks
back to ocean

turtle nest

DO NOT DISTURB

SEA TURTLE

🐢 **NEST** 🐢

**VIOLATORS SUBJECT
TO FINES AND
IMPRISONMENT**

FLORIDA LAW
CHAPTER 370

U.S. ENDANGERED
SPECIES ACT OF 1973

No person may take, possess,
disturb, mutilate, destroy, cause
to be destroyed, sell, offer for
sale, transfer, molest, or harass
any marine turtle or its nest or
eggs at any time.

No person may take, harass,
harm, pursue, hunt, shoot,
wound, kill, trap, or capture any
marine turtle, turtle nest, and/or
eggs, or attempt to engage in any
such conduct.

Upon conviction, a person may
be imprisoned for a period of up
to 60 days or fined up to $500, or
both, plus an additional penalty of
$100 for each one turtle egg
destroyed or taken.

Any person who knowingly
violates any provision of this act
may be assessed a civil penalty
up to $25,000 or a criminal
penalty up to $100,000 and up to
one year imprisonment.

SHOULD YOU WITNESS A VIOLATION, OBSERVE AN INJURED
OR STRANDED TURTLE OR MISORIENTED HATCHLINGS,
PLEASE CONTACT FWC AT

1-888-404-FWCC or ★FWC (MOBILE PHONE)

FLORIDA FISH AND WILDLIFE CONSERVATION COMMISSION
MARINE TURTLE PROTECTION PROGRAM

Warning signs help people
avoid disturbing turtle
nests and remind people of
laws protecting the nests.

Newly hatched sea turtles scurry to the water
before predators notice them. Only about 1 in
1,000 survive to become adults.

After the hatchlings leave the nest, scientists record
how many eggs hatched and how many did not survive.

Hatchlings

*Green sea turtles hatch after about 60 days. The newly hatched
sea turtles dig their way out of the nest and race to the ocean.*

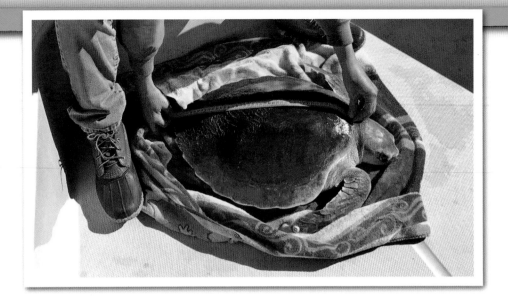

Kristen and the USGS scientists also tag and track sea turtles. Scientists temporarily capture female sea turtles on the nesting beaches. Biologists measure and tag the sea turtles. DNA samples are also taken.

Scientists have attached a tag and painted a number on the shell so the turtle is easy to identify from a distance.

After tagging the turtle, Kristen and the science team release it back to the sea.

Different kinds of tagging **techniques** help experts track daily location and movement patterns of the tagged turtles. This helps biologists understand individual and species preferences to home ranges, finding food, and nesting grounds.

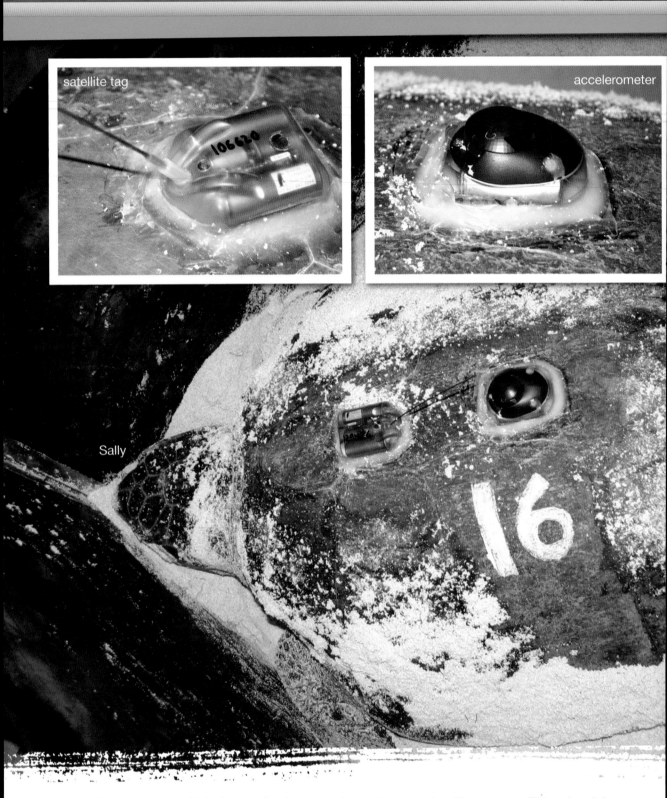

Sally, an adult female loggerhead sea turtle, was fitted with a satellite tag. Satellite tags track where the animal travels across the sea or to different depths.

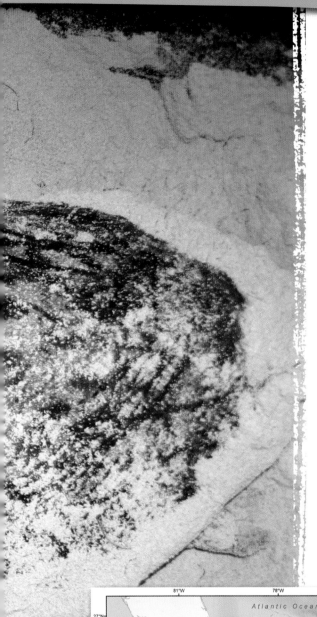

Sally was also fitted with an accelerometer. An accelerometer is a small motion detector. It uses the same technology found in smart phones and video game controllers. The accelerometer records how the animal moves, such as turning, swimming, diving, eating, or resting.

By comparing data from the satellite tags with the accelerometer tags, scientists like Kristen can create a picture of the daily lives of sea turtles. Knowing where they eat, nest, or rest lets experts know those habitats need protection for the turtles.

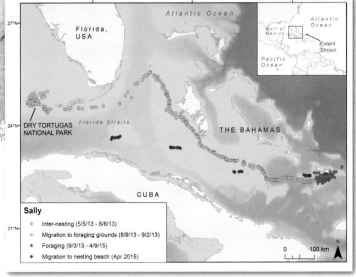

Satellite tag and accelerometer data tell scientists where Sally traveled and what she was doing.

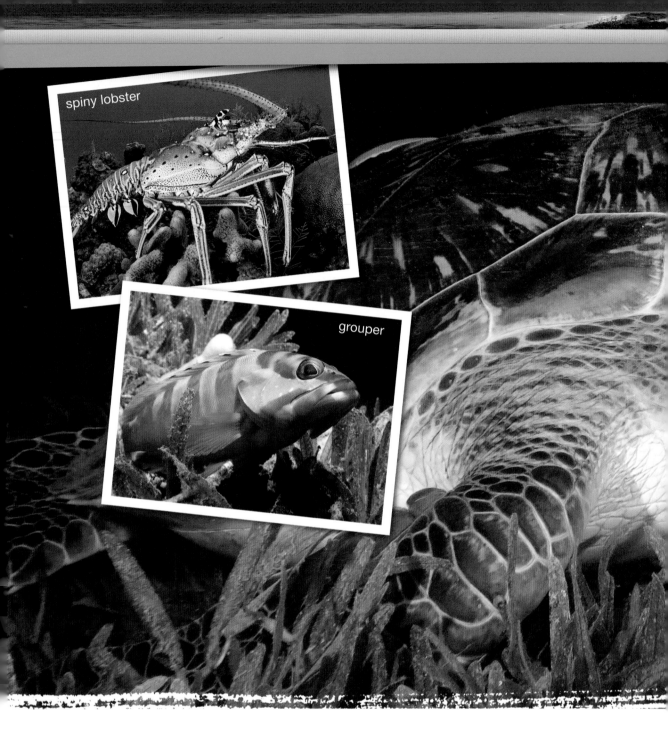

spiny lobster

grouper

Seagrass meadows are an important habitat for sea turtles. Seagrasses are plants that live in the ocean. Sea turtles, fish, and other animals rely on seagrass meadows for food and homes. Meadows are also nursery areas for spiny lobsters and fish such as grouper.

A View from Space

Seagrass is not a kind of grass. It is related to flowering plants. The name describes the long, thin leaves, which look like blades of grass. Seagrass habitats are found around the world. Some are so big they're visible from space.

Seagrass is also important for the health of coral reefs and other ocean habitats. Seagrass is a natural water filter and helps maintain water quality in the ocean. If seagrass communities are reduced or damaged, severe problems are caused to the coral reefs.

Nutrient-rich seagrass meadows filter and trap sediment from the water, helping to keep the ocean clean and clear.

Filtration System

The clear water seen throughout the Caribbean is partially due to the filtration abilities of seagrass.

Because a seagrass habitat can be damaged by human activities, scientists keep an eye on the health of the meadows. They measure the size of the meadows and count the number of seagrass species. Biologists also study the animal species within the seagrass and how they use the habitat.

Sea Turtles and Seagrass

Because green sea turtles primarily eat seagrass, maintaining healthy seagrass meadows is essential for sea turtles' survival.

The heart and soul of the Dry Tortugas ecosystem is its extensive coral reefs. Coral reefs provide food and shelter to fish, crabs, octopuses, sponges, seastars, and many other animals.

Reef Fish

There are 330 species of fish that directly rely on the reefs within the Dry Tortugas.

Corals are animals related to sea anemones and jellyfish. An individual coral, called a polyp, looks like a tiny sea anemone. It is the hard corals that create the reefs. Hard corals create rock-like **exoskeletons**.

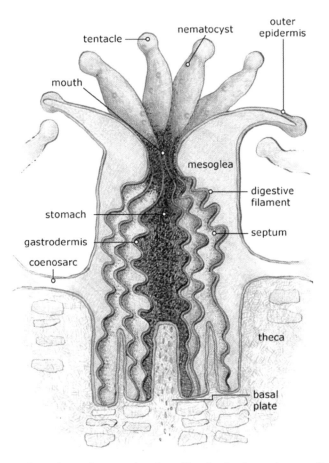

This is a cross-section view of a coral polyp. The tentacles release stinging cells (nematocysts) to capture microscopic plants and animals.

Building a Reef

Coral polyps cement themselves together to create a colony. Different colonies join together and over hundreds or even thousands of years they create a reef.

Hard corals get their beautiful coloration from **microscopic** algae called zooxanthellae (zoh-zan-THEL-ee). The zooxanthellae live within the polyp. Zooxanthellae get their energy from the sun through photosynthesis.

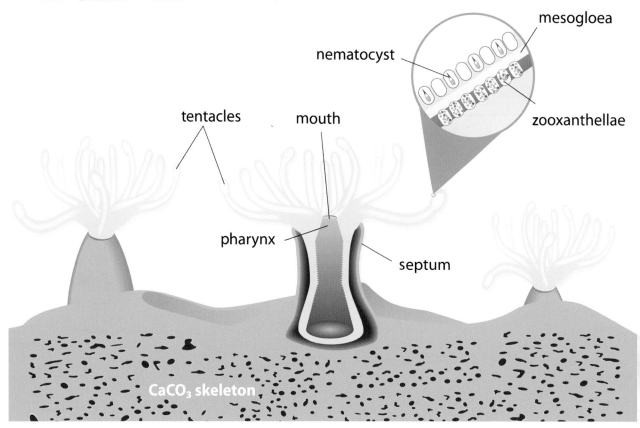

mesogloea

nematocyst

zooxanthellae

tentacles

mouth

pharynx

septum

CaCO₃ skeleton

elkhorn coral

Elkhorn coral and staghorn coral were once the most common kinds of coral in the Dry Tortugas. Both species are now threatened by disease, warming seas, hurricanes, and other factors.

Scientists scuba dive to monitor the health of staghorn coral colonies.

Creating New Colonies

Elkhorn and staghorn coral form new colonies when branches break off. The broken branch attaches to the seafloor and a new colony begins to grow. This type of reproduction allows new colonies to grow very quickly and the coral to recover after a hurricane.

Jeff Miller is a fisheries biologist with the National Park Service's South Florida/Caribbean Inventory and Monitoring Program. He routinely monitors the health and size of coral colonies. Disease, bleaching, hurricanes, water temperature variations, and human impact can affect coral reefs.

Reefs and Healthy Oceans

The health of the coral reefs also indicates the health of the ocean.

Bleaching happens when zooxanthellae in the coral polyps die. That causes the coral to turn white. Bleaching is caused by extreme temperature changes, pollution, and other environmental changes.

CORALBLEACHING

Have you ever wondered how a coral becomes bleached?

HEALTHY CORAL
1 Coral and algae depend on each other to survive.

Corals have a symbiotic relationship with microscopic algae called zooxanthellae that live in their tissues. These algae are the coral's primary food source and give them their color.

STRESSED CORAL
2 If stressed, algae leaves the coral.

When the symbiotic relationship becomes stressed due to increased ocean temperature or pollution, the algae leave the coral's tissue.

BLEACHED CORAL
3 Coral is left bleached and vulnerable.

Without the algae, the coral loses its major source of food, turns white or very pale, and is more susceptible to disease.

WHAT CAUSES CORAL BLEACHING?

Change in ocean temperature
Increased ocean temperature caused by climate change is the leading cause of coral bleaching.

Runoff and pollution
Storm generated precipitation can rapidly dilute ocean water and runoff can carry pollutants — these can bleach near-shore corals.

Overexposure to sunlight
When temperatures are high, high solar irradiance contributes to bleaching in shallow-water corals.

Extreme low tides
Exposure to the air during extreme low tides can cause bleaching in shallow corals.

NOAA's Coral Reef Conservation Program
http://coralreef.noaa.gov/

Surviving Bleaching

Coral can survive bleaching if the environment returns to normal quickly.

27

Every year, Jeff and other scientists dive to the same 100 sites within the reef. They make visual records with photos and video. They count the types and quantity of coral and other animals. They look for any signs of bleaching or diseases on the coral. Later, they compare the sites with previous years' records to look for changes or trends.

"Unfortunately, we've found there is less live coral in the Dry Tortugas than there used to be. Much of this coral loss has been caused by coral diseases, and these diseases have been more common in the past 10 years." — Jeff Miller

Many species of sharks live in or visit the Dry Tortugas. The clear reef waters give biologists a special window into the behavior of nurse sharks. Nurse sharks hunt for reef fish, lobsters, crabs, and shrimp within the coral.

Nurse sharks reach lengths up to 8 feet (2.4 meters). That's about the length of two 8-year-old children.

Scientists monitor the breeding behavior of nurse sharks.

Nurse sharks also use the Dry Tortugas to find mates and as a nursery for their young. Every summer, scientists tag nurse sharks. They also collect **genetic** material and measure and weigh the sharks. Genetic studies help scientists track how and if animals are related.

To study the sharks, scientists use different tags to track individuals. Information gathered includes movement, swimming depth, and habitat preferences. Scientists have discovered that some sharks spend most of their lives in the Dry Tortugas. Others return seasonally to find mates or give birth.

Tagged Sharks

Every year, scientists recapture about a quarter of the sharks tagged the previous year. Scientists update the records on those animals.

Like other seabirds, roseate terns find their food in the sea. They look for fish from the air and then dive into the water to catch them.

Ricardo Zambrano, Wildlife Biologist with the Florida Fish and Wildlife Conservation Commission, studies roseate terns. These seabirds nest on isolated beaches, like the Dry Tortuga Islands.

A real roseate tern checks out a decoy tern.

The return of roseate terns to the Dry Tortugas is a result of Ricardo's research and conservation efforts.

By 2006, the terns had not nested in the Dry Tortugas for more than 30 years. Ricardo and the science team used a few tricks to encourage the birds' return. They used decoys in the shape and colors of the terns. They also broadcast the birds' calls on outdoor speakers 24 hours a day. It worked. Roseate terns have been nesting in the Dry Tortugas ever since.

Decoys

Wooden decoys make a beach look populated with terns and that makes the area inviting to birds looking for a nesting site.

Chick Check

Every year, Ricardo counts the number of adults, nests, and chicks. They also put small numbered bands on the chicks' legs. Scientists can identify the chicks as they grow up and where they travel.

CHAPTER THREE
CONSERVATION

Discoveries made in the Dry Tortugas help biologists make decisions on how to protect wildlife and manage **commercial** and recreational fishing within the park.

Experts like Kristen, Jeff, Ricardo, and other scientists also study long-term trends in animal population sizes and changes. The data they collect helps track the health of the coral reef system and the ocean.

Restricting and limiting the size and quantity of boats allowed to enter the Dry Tortugas helps preserve the habitat.

THREATS TO CORAL REEFS OVERFISHING

Coral reef fish are a significant food source for over a billion people worldwide. Many coastal and island communities depend on coral reef fisheries for their economic, social, and cultural benefits.

BUT too much of a good thing can be bad for coral reefs.

FISHING NURSERIES
Nearshore habitats serve as nurseries for many fish. Catching young fish in nets removes them before they can help replenish the population.

MARINE DEBRIS
Traps set too close to reefs and marine debris, such as ghost traps, lost nets, monofilament, and lines, can damage coral reefs, which take a long time to recover.

INDISCRIMINATE FISHING
Use of non-selective gears, like nets and traps, often removes more herbivorous fishes. These fish eat algae and help keep the ecosystem in balance.

FISHING SPAWNING AGGREGATIONS
Some species gather in large numbers at predictable times and locations to mate. Spawning aggregations are particularly vulnerable to overfishing.

FISHING TOO MANY BIG FISH
Large fish produce more young that are likely to survive to adulthood. Their absence means fish populations dwindle over time.

HOW YOU CAN HELP

 Educate yourself on local fishing rules and regulations. Your state fishery agency or bait and tackle shop can help you learn more.

 Make sustainable seafood choices. Learn more at www.FishWatch.com.

 Only take what you need. Catch and release fish that you don't plan to eat.

 Be a responsible aquarium owner. Know where your fish come from and DO NOT release unwanted fish into the wild.

Throughout the world, overfishing—taking too many fish and other ocean animals—is having disastrous effects on animal populations and habitats.

Knowing what areas animals need to find food and shelter, or for nesting and nurseries helps scientists recommend protective measures to lawmakers. This kind of information led to the creation of a marine **reserve** within the Dry Tortugas.

Today, nearly half of Dry Tortugas National Park has been set aside to protect animals and their habitats. This area is called the Research Natural Area (RNA).

The RNA provides sanctuary for plant and animal species and helps protect animals affected by overfishing and habitat loss in the Gulf of Mexico.

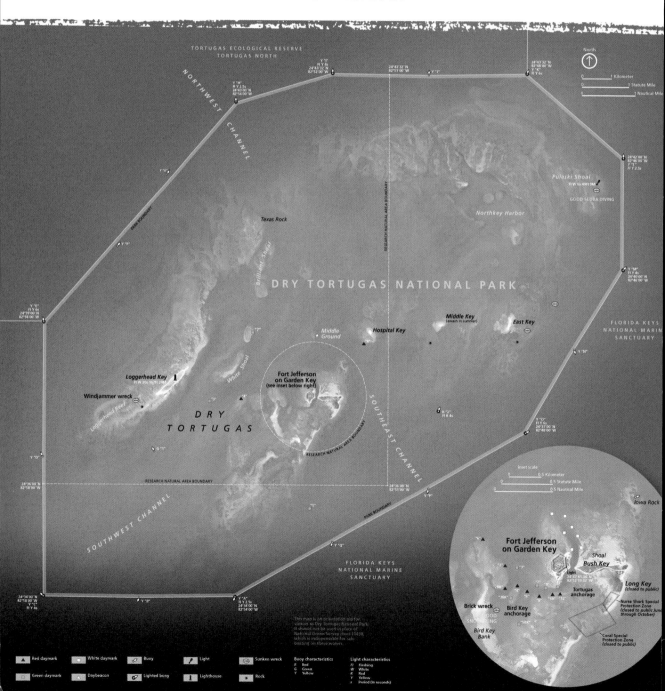

Commercial and recreational fishing are not allowed within the RNA boundary. This means animals within the RNA are protected from hunting or harassment.

Some fishing is permitted within specified areas of the Dry Tortugas and is closely monitored to keep fish populations healthy.

Mooring buoys are used instead of anchors to protect the ocean floor and allow a boat to remain in one area.

Anchors cause damage to coral, seagrass, and other life on the ocean floor.

Habitats are damaged when boat anchors drag across reefs and seagrass meadows. Within the RNA, there are strict restrictions on where boats can drop anchor in order to protect delicate habitats.

Dr. Jerald Ault led a team that conducted 3,000 dives over nine years to count fish populations throughout the Dry Tortugas. The goal was to compare areas where fishing was allowed to areas where it was not. Research showed that many fish populations had increased within the protected area.

Scientists use tablets in waterproof cases and paper that doesn't fall apart in the water to make notes during dives.

"We don't manage fish, turtles, or coral, we manage human activities that affect those animals." — Kristen Hart

Results from Kristen's USGS studies show that green sea turtles benefit from the RNA by nesting and feeding within the protected area.

Ecosystems don't start and end within human-created boundaries. In order to protect animals from overfishing or habitat **destruction**, it is important to understand how animals move between ecosystems. Scientists from the Dry Tortugas National Park work closely with researchers in connected ecosystems and national parks through the South Florida/Caribbean Network.

FLORIDA

Big Cypress NP

Everglades NP   Biscayne NP

Dry Tortugas NP

CAYMAN ISLANDS

CUBA

JAMA

Park System

- *There are seven parks in the South Florida/ Caribbean Inventory and Monitoring Network.*

E BAHAMAS

TURKS
AND CAICOS
ISLANDS

HAITI

British
Virgin
Islands

DOMINICAN
REPUBLIC

PUERTO
RICO

Virgin
Islands

Understanding the relationship between the animals and habitat in Dry Tortugas National Park helps scientists protect them and predict how changes in habitat can affect species. Many plants and animals are good **indicators** of an ecosystem's health.

A sea turtle relaxes while cleaner fish remove dead skin and parasites from its skin.

NEIGHBORHOOD SCIENCE

Discover how scientists monitor a large habitat by gathering data from small areas. Think like a scientist to discover what plants and animals live near you.

Supplies

- 8 feet (2.5 meters) yarn or twine
- paper
- pencil
- color pencils or crayons
- clipboard (optional)
- camera (optional)

Directions:

1. Tie the ends of the twine together. This will allow you to collect data at different places within the same size space.

2. Choose a location to gather your data: a local park, playground, or your own yard. Make a list of what you expect to find.

3. At the location, carefully set down the twine circle to create the small study area.

4. Write down the type of plants, animals (probably insects), or other things you see within the study area. Draw the area, noting where you see each thing. Take photos to document what you see.

5. Repeat steps 3 and 4 in a couple different places.

6. Sort and classify the data you collected. Figure out the best way to represent your data, such as graphs, writing, or art.

7. Using what you found in the small study areas, estimate how many of each plant or animal might be found within the park, playground, or your yard.

8. Did you find what you expected? Why or why not? Looking at the data, what questions do you have? How can you answer them?

Glossary

abundant (uh-BUHN-duhnt): a large quantity of something

commercial (kuh-MUR-shuhl): things that can be bought or sold

destruction (di-STRUK-tiv): to destroy or damage something

endangered (en-DAYN-jurd): describes a plant or animal group that is getting too small and could become extinct if it continues to get smaller

exoskeletons (ek-so-SKEL-o-tuhns): the hard outside coverings that protect animals, such as the shell of a crab or the rock-like covering of coral that creates the reef

genetic (juh-NET-ik): having to do with how an animal is related to another

indicators (IN-di-kay-turz): information that points to or hints of some other information

microscopic (mye-kruh-SKAH-pik): something that is so small it cannot be seen without special equipment like a microscope

monitor (MAH-ni-tur): to watch or check on over time

pristine (PRISS-teen): unspoiled, in perfect condition or never used

reserve (ri-ZURV): natural area set aside to protect the entire ecosystem including the plants, animals, and landscape

techniques (tek-NEEKS): ways of doing things that require practice, training, or skill

Index

Show What You Know

1. How many islands are in Dry Tortugas National Park?

2. How is a coral reef created?

3. Why is it important to protect seabed meadows?

4. What are two kinds of hard coral?

5. How did scientists trick roseate terns into nesting in Dry Tortugas National Park?

Further Reading

National Parks Guide USA Centennial Edition: The Most Amazing Sights, Scenes, and Cool Activities from Coast to Coast! National Geographic Children's Books, 2016.

Young, Karen Romano. *Mission: Sea Turtle Rescue: All About Sea Turtles and How to Save Them.* National Geographic Children's Books, 2015.

Wicks, Maris. *Science Comics: Coral Reefs: Cities of the Ocean.* First Second, 2016.

About the Author

Ruth A. Musgrave talks with turtles, has sea stars in her eyes, and counts sharks among her most trusted friends. Ruth is also an award-winning author of hundreds of articles about animals and more than 19 books including *Mission Shark Rescue: All About Sharks and How To Save Them* (National Geographic Kids, 2016). Ruth is also a naturalist and lucky hitchhiker on ocean research cruises including diving to the deep sea. Find out more at www.ruthamusgrave.com.

www.rourkeeducationalmedia.com

PHOTO CREDITS: Cover foreground photo © Rich Carey, card with paper clip art © beths; title page and background photo on cover © T. Anderson Photography; contents page © Joe Quinn; page 4-5 long stretch beach © Ovidiu Hrubaru; page 6-7 turtle © Alita Xander; pages 8-9 loggerhead © Joe Quinn, green turtle © Been there YB, leatherback © IrinaK, page 9 people on beach © Serenethos, condos © Kristi Blokhin; beach photo page 10-11 © Benny Marty, map page 11 © boreala; page 12 large photo © ymgerman, turtle eggs © David Evison, returning to sea © Simon Eeman, page 13 large photo © Kevin Wells Photography, baby turtle © rujithai; page 18-19 turtle and grouper shots © Laura Dinraths, spiny lobster © Ethan Daniels; page 20-21 © Richard Whitcombe, page 21 © Ethan Daniels; page 24-25 illustration © Designua, elk horn coral photo © Edwin van Wier; page 27 inset photo © MYP Studio; page 29 © Ethan Daniels, page 29 top photo © tali de pablos; page 35 © Kaisha Morse; page 36-37 coral reef © Huw_Thomas06; page 39 © Varina C; page 40 mooring buoy © ultrapok, anchor photo © C Levers; page 42-43 map © Sira Anamwong. All images from Shutterstock.com except: page 5 bottom inset photo courtesy of NPS; page 6 map courtesy of Library of Congress, Ponce de Leon public domain; page 8 Kemp's ridley turtle by Jereme Phillips, USFWS. https://creativecommons.org/licenses/by/2.0/ , hawksbill courtesy Caroline Rogers, U.S. Geological Survey. Public domain; page 10 inset turtle nest and portrait courtesy Kristen Hart USGS, page 13 sign © Ianaré Sévi https://creativecommons.org/licenses/by-sa/3.0/deed.en ; page 14 measuring turtle courtesy of USGS, bottom photo and all photos on pages 16-17 courtesy of Kristen Hart, USGS, page 15 courtesy of Andrew Crowder, USGS; page 20 illustration © Craig Lopetz, page 21 green turtle in seagrass © Johninpix | Dreamstime.com; page 22-23 coral photos courtesy of NPS, coral diagram courtesy of NOAA; page 25 courtesy of USGS; page 26: monitoring size and health of coral images courtesy of NPS, p27 coral bleaching chart courtesy of NOAA; page 28 biologist and coral courtesy of NPS; Page 30 © (c) Nicolas Voisin | Dreamstime.com, page 31 © Harold Wes Pratt; page 32-33 main photo courtesy of U.S. Fish and Wildlife Service Southeast Region, https://creativecommons.org/licenses/by/2.0/ , page 33 decoy photos and all photos on page 34 except egg photo courtesy of Ricardo Zambrano, Regional Biologist, Florida Fish and Wildlife Conservation Commission, page 34 egg photos courtesy of NPS; page 36 infographic courtesy of NOAA; page 38 courtesy of NPS; page 40 inset photo courtesy of NOAA, page 41 courtesy of NPS; page 43 turtle courtesy of Kaare Iverson, USGS; page 44 courtesy of Thierry Work, USGS, author photo © A. Buchholz

Edited by: Keli Sipperley

Produced by Blue Door Education for Rourke Educational Media. Cover design and page layout by: Nicola Stratford

Dry Tortugas / Ruth A. Musgrave
  (Natural Laboratories: Scientists in National Parks)
  ISBN 978-1-64369-025-4 (hard cover)
  ISBN 978-1-64369-114-5 (soft cover)
  ISBN 978-1-64369-172-5 (e-Book)
Library of Congress Control Number: 2018956044

Printed in the United States of America, North Mankato, Minnesota